What to watch on Amazon Fire Stick?

There are number of things you can watch on Amazon Fire Stick. You can not only watch different content but you can also listen to music on Amazon fire Stick. Amazon has instant play music and instant videos.

In this book you will discover a step by step approach to unlock your fire stick so let's begin.

Special Note: it is not illegal to unlock Fire Stick, however the usage of Fire Stick without paying for the content is illegal.

1. Go to the Home page first.

2.   Now scroll to the bottom of the page and you will find "Settings" options.

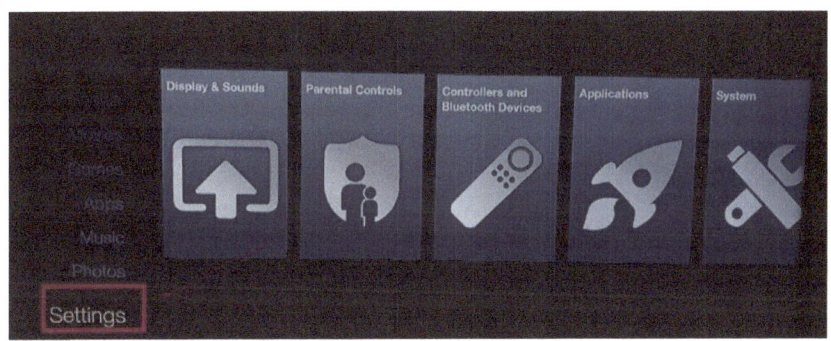

3.   Now scroll to the right of the screen and you will see "System" options.

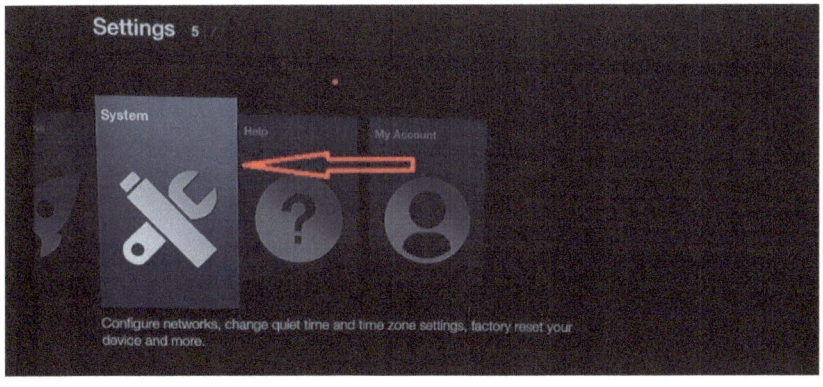

4. Click on the "system" options and you should see a "Developer Options".

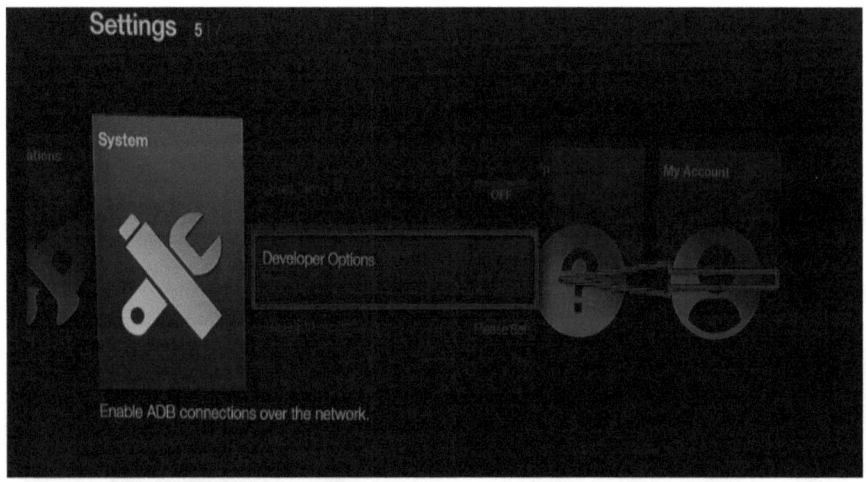

5. After clicking on "developer Options you should see "ADB debugging", simply turn it ON.

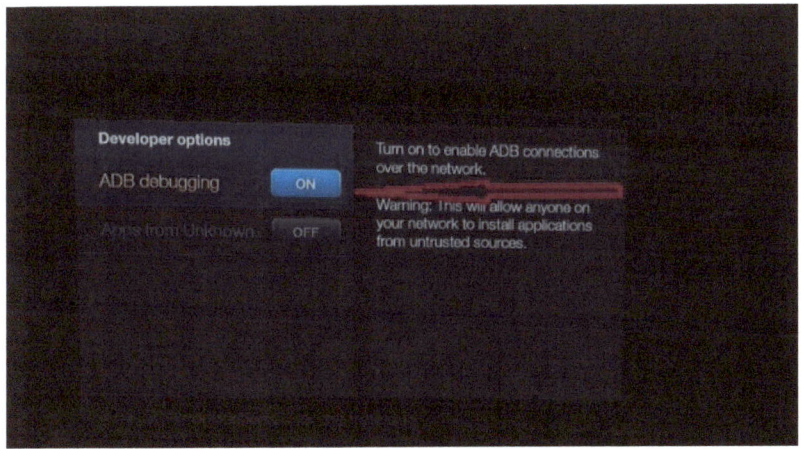

6. now go to the "Apps from Unknown sources" and turn it on, upon tuning it on.

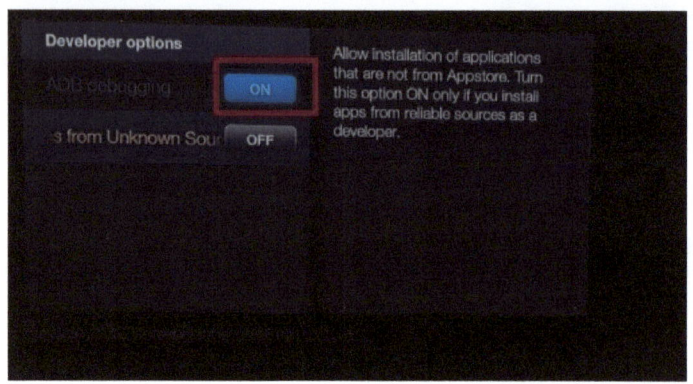

7. Upon turning it on you will see a warning message, simply click OK

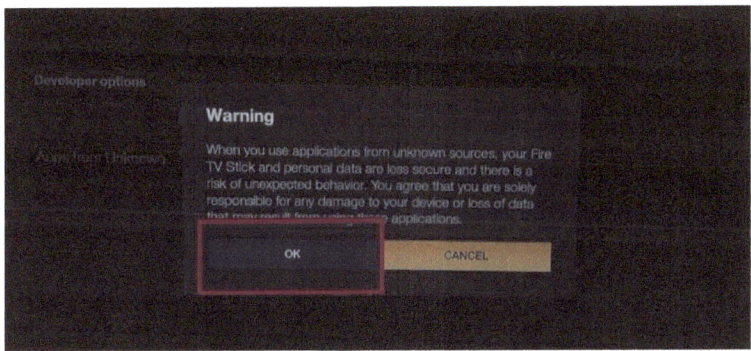

8. Once you have completed the above steps simply click "Home" button on your remote and you will be taken to the home screen. Now scroll up to "Search".

Text Search

A B C D E F G H I J K L M
N O P Q R S T U V W X Y Z
1 2 3 4 5 6 7 8 9 0

9. Now simply type "Es Explore". Or select from the drop down menu, (you will see drop down list once you start typing).

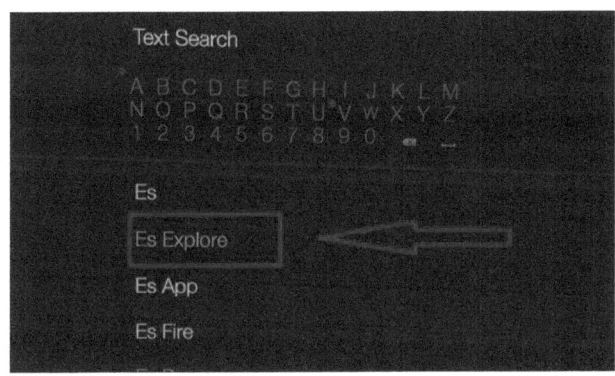

10. Now you will be presented a screen similar like this simply scroll to "Apps & Games".

11. Now you will see Es Explore page.

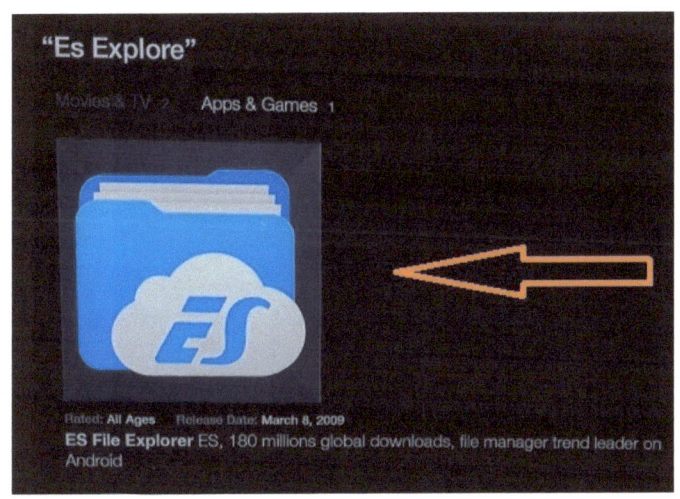

12. Now click on the icon and you will be inside Es Explore, you will see an potion of "Download" simply click on "Download."

13. Once you have clicked download button you see the status on your screen it will automaticlally download and install.

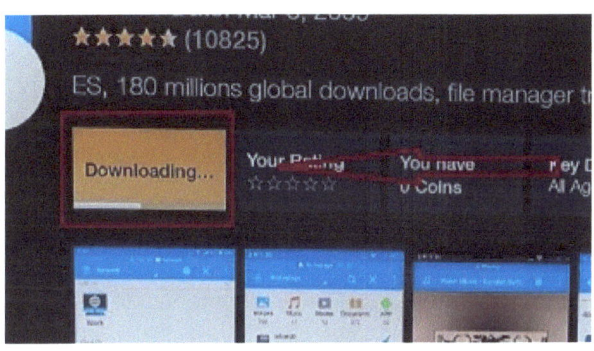

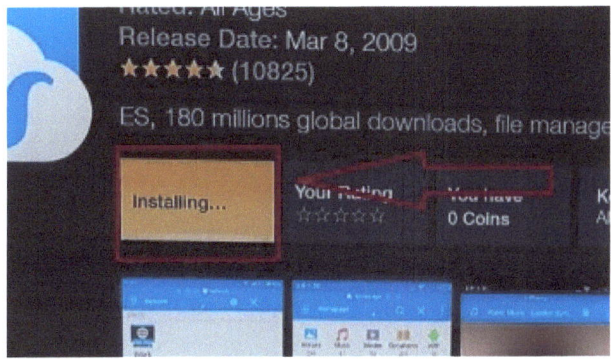

14. Once the install is finished you will see "Open" button.

15. Now simply click on Open button and after few seconds you will see a screen.

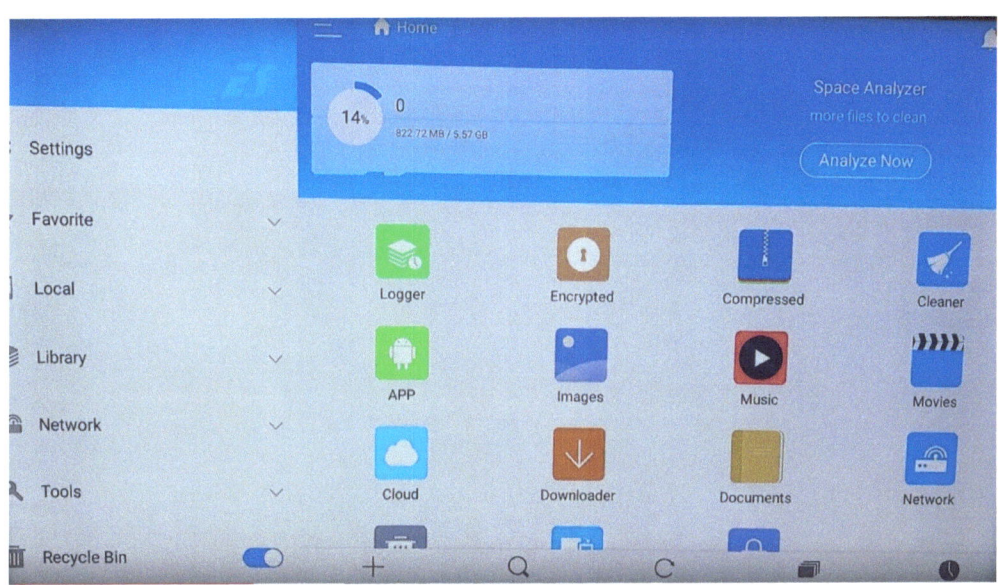

16. On the left side simply scroll to "Favorite" option.

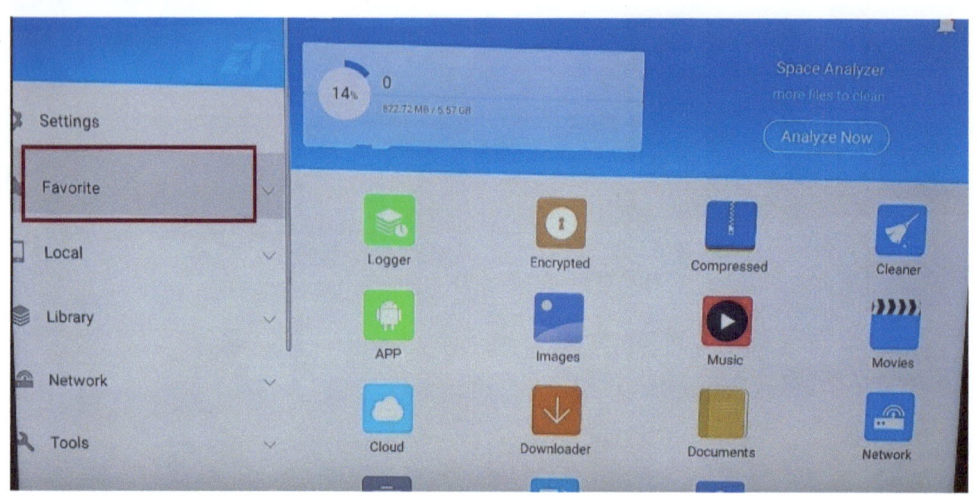

17. Underneath "Favorite" you will see "Add" option simply click on Add.

18. Now you will see a screen called "Add to favorite."

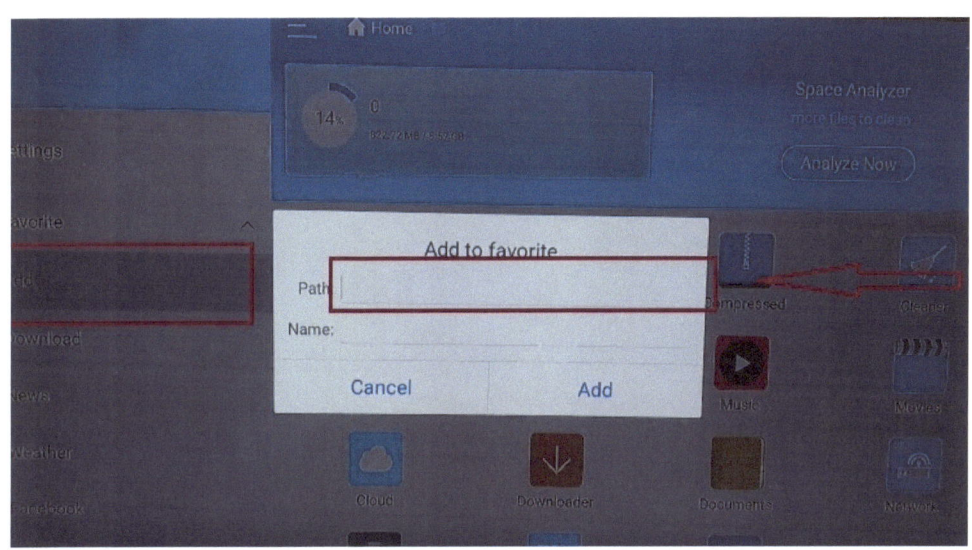

19. Cliak on path and simply type the following :
https:\\kodi.tv\download

and simply click on "Next" button.

20. Now in the box you will need to enter a desired name your choice. Simply put any name you like I have put 'kodi'. Now simply click on "Next" button.

21. Now you will see a screen like this, simply click on "Add."

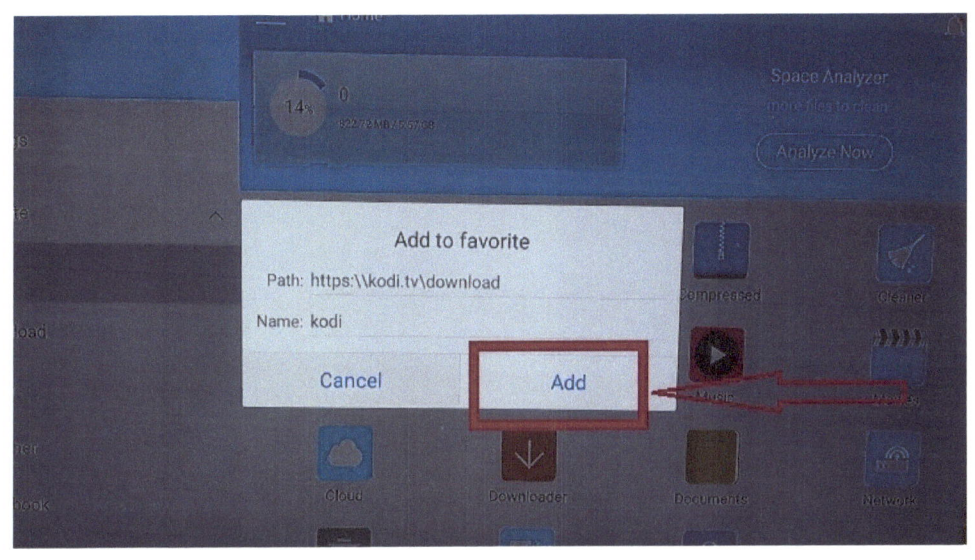

22. Once done you will see a
message like "Bookmark kodi
is created successfully."

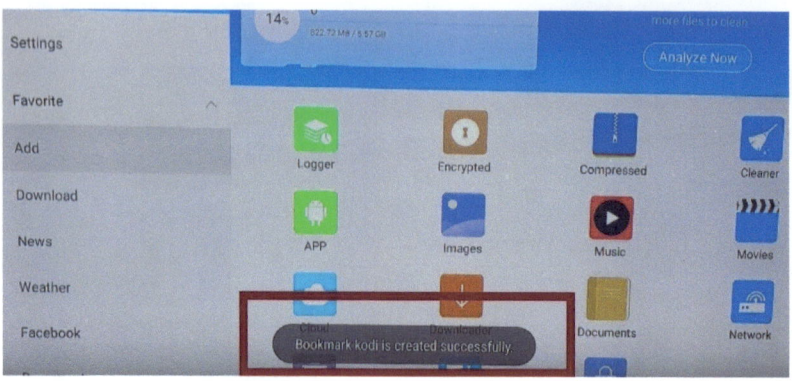

23. Now scroll bottom from
the left hand side menu and
you will find your named file,
in this case 'kodi.'

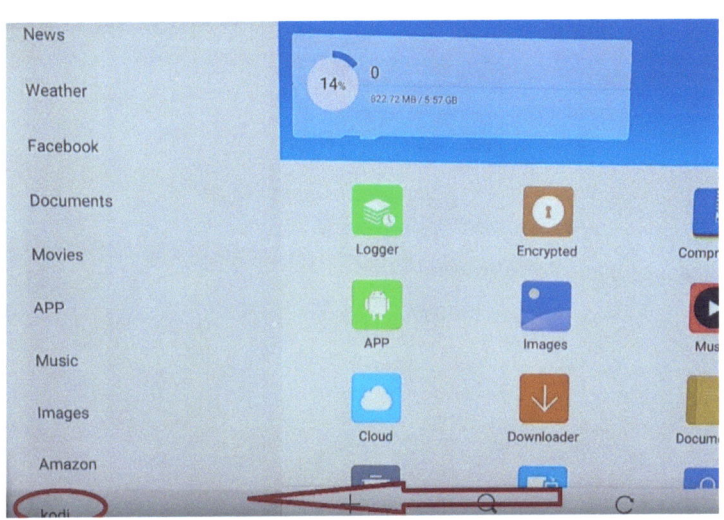

24. Now you will be presented with a screen like this

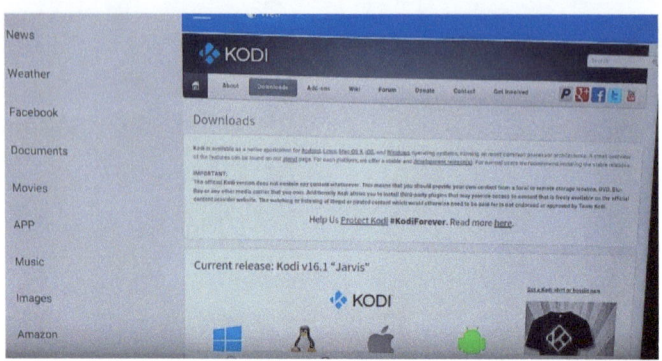

25. Now you will need to use your Fire stick remote to navigate this screen. You need to go to the icon as shown in the figure.

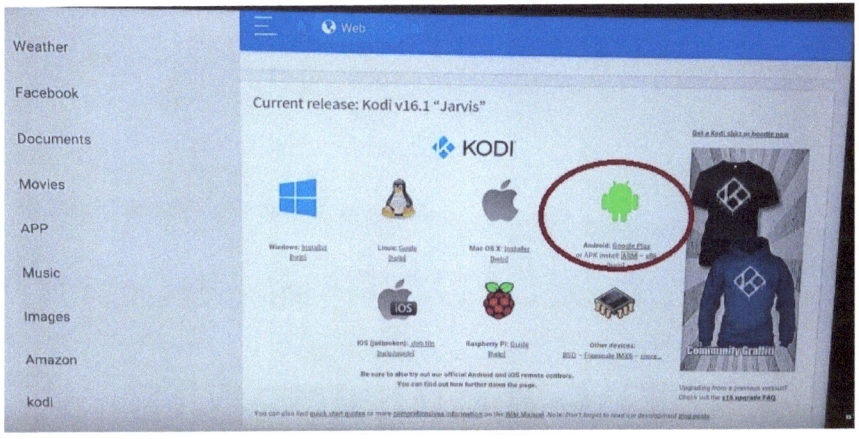

26. Now click on the option highlighted "ARM"

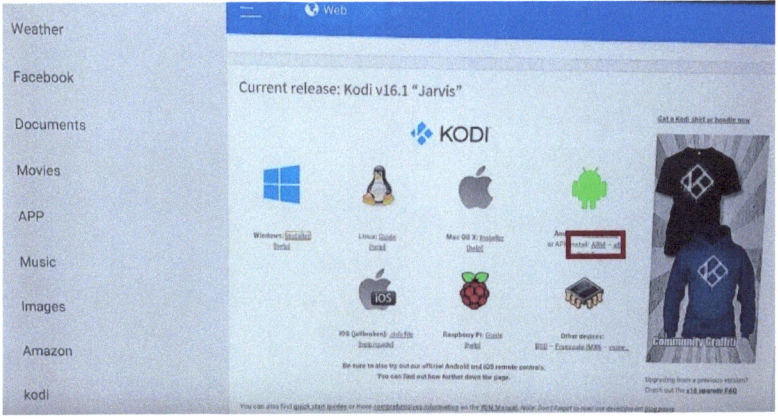

27. Now wait for few seconds to load the download page.

28. After downloading you will be asked to "Open folder or "Open file". You need to select "Open file".

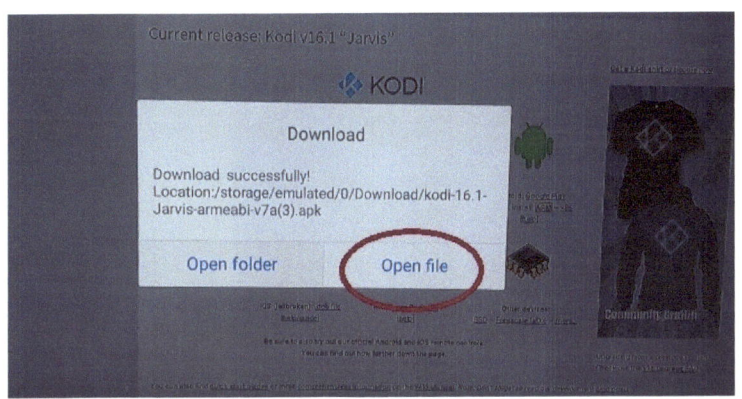

29. Now you will see a screen like this, simply cliak "Install."

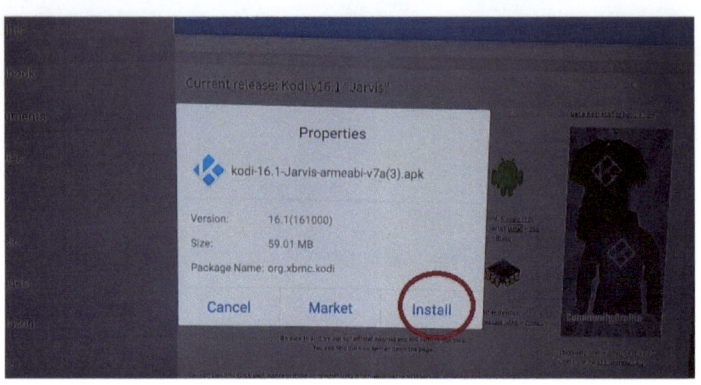

30. Now you will see a careen like this which will say "Installing…"

31. After finish installing you will see a screen like this simply click on OPEN.

Kodi

✓ App installed.

DONE   OPEN

32. After few minutes you will be presented a careen like this.

33. Congrats you have successfully unlocked your Firestick.

www.ingramcontent.com/pod-product-compliance
Lightning Source LLC
Chambersburg PA
CBHW041133200526
45172CB00018B/310